Unit 12

Using Scientific Notation

Developed by the Center for Occupational Research and Development and sponsored by a consortium of State Vocational Education Agencies with the cooperation and support of mathematics educators.

Vocational occupations icon design courtesy of the Vocational Education Services Project directed by Dale Law and John Smith of the University of Illinois.

Published and distributed by:
CORD Communications
324 Kelly Drive
Waco, Texas 76710
817-776-1822 FAX 817-776-3906

Printed in USA April 94

Library of Congress Catalog Card Number: 87-072538
ISBN 1-55502-296-0 (Applied Mathematics)
ISBN 1-55502-315-0 (Unit 12, Using Scientific Notation)

PREFACE

Applied Mathematics contains video programs, laboratory activities and problem-solving exercises. Each part has been chosen to help you understand the mathematics you need to work and live in a technical world. Most importantly, each part has been designed to make mathematics more useful and meaningful for you—and reduce some of the "math anxiety" we all feel at one time or another.

Applied Mathematics contains some of the important mathematics you'll need to be a productive member of today's workforce. The ideas you'll learn will help you understand:

- numbers, decimals, fractions and percents;
- shapes and sizes;
- how to handle equations and formulas;
- how to work with angles and triangles;
- how to estimate answers and solve problems; and
- how to describe the behavior of large populations of things.

Each of these skills will help you do your job better—and help you advance to the next job when the time comes.

Each unit of **Applied Mathematics** begins with a video program. The video tells you about the mathematics skills you'll be studying and introduces you to real people who use these skills in their everyday lives. Following this, you'll concentrate on learning the mathematics skills.

Next, you'll have a chance to practice what you've learned. You'll do this in laboratory activities that involve measurement and problem solving. You'll apply the math skills to practical problems. These problems are the kind that people have to solve every day—in restaurants, on the farm, in a factory, in a business office, in a hospital, in a laboratory, or at home.

You'll have many opportunities to practice what you've learned, so don't be discouraged if it all doesn't seem too clear the first time around. You'll learn how to use mathematics, little by little, problem by problem. And in the process, you may find that using mathematics can be fun.

The CORD Project Staff

Table of Contents

LEARNING THE SKILLS

PRACTICING THE SKILLS

REFERENCE MATERIALS

Using Scientific Notation

How to write numbers in scientific notation and solve problems using numbers in scientific notation

Prerequisites

This unit builds on the skills taught in:

Unit 1: *Learning Problem-solving Techniques*
Unit 2: *Estimating Answers*
Unit 3: *Measuring in English and Metric Units*
Unit 11: *Using Signed Numbers and Vectors*

To Master This Unit

Read the text and answer all the questions. Complete the assigned exercises and activities. Work the problems on the unit test at a satisfactory level.

Unit Objectives

Working through this unit helps you learn how to:

1. Write large and small numbers in power-of-ten notation.

2. Read and write numbers in scientific notation.

3. Enter numbers written in scientific notation into a calculator and read answers in scientific notation displayed by a calculator.

4. Combine numbers written in scientific notation to solve problems.

Learning Path

1. Read the "Introduction" section of this unit.
2. Watch the video and take part in the class discussions.
3. As you read the text, think about what the words say, and answer any questions that the text asks. Follow the signals that help you learn, including the ones that tell you to work problems and write on your own paper.
4. Do the assigned math lab activity.
5. Complete the assigned exercises.
6. Measure your progress by taking the unit test.

Some Signals to Help You Learn

The following signals help you know what to do as you read the text:

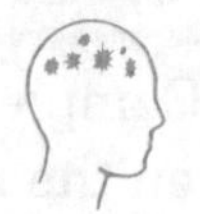

Think this through. Spend a little extra time on this idea.

Write your answer on your paper.

Carry out the calculations.

Learn this key rule or definition.

Estimate and ask yourself if this answer makes sense.

Compare your answer to the given one and make any needed changes.

LEARNING THE SKILLS

INTRODUCTION

What size numbers do you work with most often? You may seldom use numbers over 10,000, and most of the time you probably use numbers that are between 0.1 and 100. You may read in newspapers about budgets in the millions—or even billions of dollars. And you may hear of measurements made to *ten thousandths of an inch*. But aside from such occasions, you may not often write or read large numbers.

Nevertheless, many people do use *very large* and *very small* numbers every day on the job. For example, people who work in the space industry, the world bank or national defense often use large numbers. Persons working in a biology lab or a center for disease control may work with human cells and viruses—and use very small numbers. As shown in Figure 12-1, astronomers look through telescopes to study stars and planets that are great distances from the earth. And biologists use microscopes to examine objects much smaller than you can see with the naked eye.

Figure 12-1
Very large and very small numbers

Let's look at other examples of very large and very small numbers:

- An ounce of gold contains approximately
  86,700,000,000,000,000,000,000 atoms.
- One atom of gold has a mass of
  0.000000000000000000000327 grams.
- The average volume of an atom of gold is
  0.00000000000000000000001695 cubic centimeters.
- The weight of the earth is about
  6,600,000,000,000,000,000,000 tons.
- A single human red blood cell is about 0.000007 meters in diameter.
- There are about 5,000,000 human red blood cells in one cubic millimeter of blood.

Figure 12-2 shows some of the instruments used in the space industry and in laboratories as people work with very large or very small numbers.

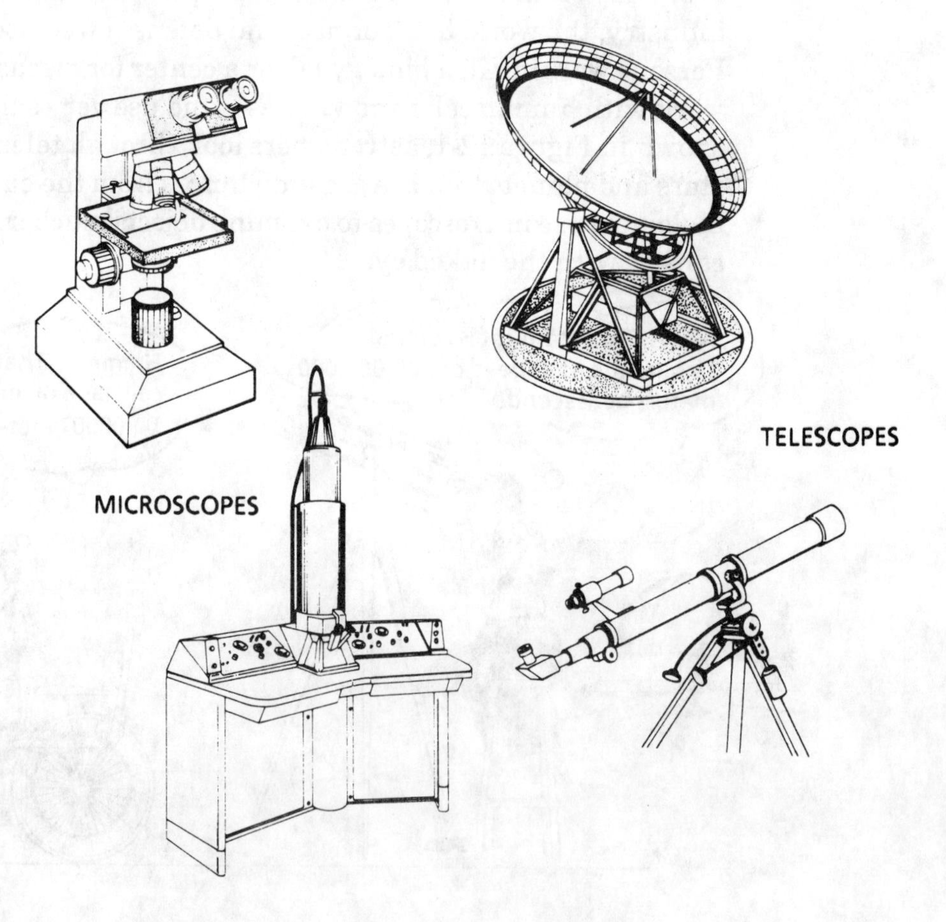

Figure 12-2
Instruments used to explore

Why scientific notation is needed

Suppose you had to calculate the number of atoms in 3.45 ounces of gold. To work this problem, you would have to multiply 86,700,000,000,000,000,000,000 atoms (the number of atoms in a single ounce) by 3.45 ounces. The large number has twenty zeros! What would happen if you tried to put this large number in your calculator?

Try pressing keys on your calculator to enter the number 86,700,000,000,000,000,000,000. On most calculators, the calculator refuses to accept more than 8 or 9 digits and this number has 23 digits—and 20 zeros! So you can't use your calculator to work the problem. And multiplying it out "longhand" is no fun. (Try it!) Are you stuck?

No! There *is* a way to work this problem with your calculator. Long before the calculator was invented, people needed to work problems with very small and very large numbers, so they invented a special way of writing those numbers. The special way of writing large and small numbers that have many zeros—without having to write down all the zeros—is called **scientific notation**. As you'll soon see, that large number of atoms in a single ounce

86,700,000,000,000,000,000,000

can be written in scientific notation as 8.67×10^{22}. The first part is just a **number between one and ten**. The last part, 10^{22}, is a **power of ten**. These are the two parts of a number written in *scientific notation*.

In this unit, you practice writing numbers in scientific notation and you use scientific notation to work problems. After you learn to write numbers in scientific notation, you will be able to use your calculator to work with numbers that have *more* than 8 or 9 digits.

As you watch the video for this unit, notice how people at work use very small and very large numbers. Notice also how they write these numbers.

POWERS OF TEN

If the two numbers shown in Figure 12-1—300,000,000 meters/sec and 0.000007 meters—had been written in scientific notation, they would have appeared as 3×10^8 m/sec and 7×10^{-6} m. The last part of each number—the 10^8 and 10^{-6}—is called a **power of ten**.

A *power of ten* is nothing more than the number 10 (called the base) raised to a power. The power (or exponent) is the number up and to the right of 10.

Before you can rewrite large and small numbers in scientific notation, you have to know how to write numbers that are multiples of 10—or divisible by 10—as **powers of 10**. Let's see how that's done.

Positive powers of ten

When you write a number such as 1000; 0.001; 100,000; or 0.00001 as a *power of ten*, you write the base 10 with an appropriate exponent. For example:

$$1000 = 10^3 \; ; \quad \text{base 10, exponent 3}$$
$$0.001 = 10^{-3} \; ; \quad \text{base 10, exponent} -3$$
$$100,000 = 10^5 \; ; \quad \text{base 10, exponent 5}$$
$$0.00001 = 10^{-5} \; ; \quad \text{base 10, exponent} -5$$

In each case, the *positive exponent* tells you how many times to write down the factor 10 and multiply to get the number. The *negative exponent* tells you how many times to write down the factor $^1/_{10}$ and multiply to get the number. You can see how this works for positive powers of ten by studying the following example:

Example 1: *Changing positive powers of ten to numbers*

Copy the following table of numbers on your paper:

$$10^4 = 10 \times 10 \times 10 \times 10 = 10,000$$
$$10^3 = 10 \times 10 \times 10 = 1,000$$
$$10^2 = 10 \times 10 = 100$$
$$10^1 = 10 = 10$$

Is there a pattern you follow as you change each power of ten to a whole number? Are the following statements true?

- For 10^4, you multiply 4 tens together and obtain an answer (10,000) that has 4 zeros after the 1.
- For 10^3, you multiply 3 tens together and obtain an answer (1,000) that has 3 zeros after the 1.
- For 10^2, you multiply 2 tens together and obtain an answer (100) that has 2 zeros after the 1—and so on.

There is a pattern! You can see that the **positive** exponent in a power of ten tells you the number of tens you multiply together to get the equivalent answer—the whole number.

Check your understanding of this table and the pattern by working through the Study Activity.

Study Activity:

a. Make a table for 10^5 and 10^6 similar to the one above. Then write sentences like those above. The sentence should relate the exponent to the number of tens multiplied together and to the number of zeros that follow 1 in the answer.

b. Then do the problem "backward." First write down the whole number and then find the power of ten it's equal to. The first line in the following table is done to get you started. Finish the other two lines.

$$\underbrace{1{,}000{,}000}_{\text{6 zeros}} = \underbrace{10 \times 10 \times 10 \times 10 \times 10 \times 10}_{\text{6 tens multiplied together}} = 10^6 \text{ (exponent is 6)}$$

10,000 =

100 =

Negative powers of ten

Now follow a similar plan to learn the meaning of negative powers of ten. Remember, the negative exponent tells you how many times to write down the factor $^1/_{10}$ and multiply to get the number.

Example 2: ***Changing negative powers of ten to numbers***

Copy the table given below on your paper and try to find a pattern.

$$10^{-4} = \frac{1}{10} \times \frac{1}{10} \times \frac{1}{10} \times \frac{1}{10} = \frac{1}{10{,}000} = 0.0001$$

$$10^{-3} = \frac{1}{10} \times \frac{1}{10} \times \frac{1}{10} = \frac{1}{1{,}000} = 0.001$$

$$10^{-2} = \frac{1}{10} \times \frac{1}{10} = \frac{1}{100} = 0.01$$

$$10^{-1} = \frac{1}{10} = 0.1$$

Is there a pattern as you change each power of ten to a number?

- For 10^{-4}, you write down the factor $^{1}/_{10}$ **four** times and multiply to get the answer 0.0001. The answer has **four** places to the right of the decimal point—three zeros and a 1.
- For 10^{-3}, you write down the factor $^{1}/_{10}$ **three** times and multiply to get the answer 0.001. The answer has **three** places to the right of the decimal point—two zeros and a 1.
- For 10^{-2}, you write down the factor $^{1}/_{10}$ **two** times and multiply to get the answer 0.01. The answer has **two** places to the right of the decimal point—a zero and a 1.

Again, there is a pattern! You can see that the number associated with the negative exponent (the 4 in – 4, for example) tells you the *number of times* you write down the factor $^{1}/_{10}$ and multiply to get the answer—and the *number of places* in the answer that are to the right of the decimal point.

Check your understanding of the table and the pattern for negative powers of ten by working through the following Study Activity.

Study Activity:

a. Make a table for 10^{-5} and 10^{-6} similar to the one above. Then write sentences that relate the exponent, the number of times $^{1}/_{10}$ is used as a factor, and the number of places in the answer to the right of the decimal point.

b. Then do the problem "backward." First write down the decimal number. Then write the power of ten it equals. The first line is done to get you started. Complete the others.

$$\overbrace{0.00001}^{\text{5 places}} = \overbrace{\frac{1}{10} \times \frac{1}{10} \times \frac{1}{10} \times \frac{1}{10} \times \frac{1}{10}}^{\frac{1}{10}\text{ five times}} = 10^{-5} \quad \text{(exponent is } -5\text{)}$$

$0.0001 =$

$0.001 =$

$0.01 =$

A detour—figuring out what 10^0 means

You have written the numbers for powers of ten, such as 10^6, 10^5, 10^4 ..., all the way down to 10^{-5} and 10^{-6}. But you passed over one particular power of ten that belongs in this sequence—10^0, or "ten to the zero power." What could 10^0 be equal to? Our previous way of writing down 10 a certain number of times and multiplying doesn't work here—for "*write down ten zero times and multiply*" doesn't make sense. Perhaps by writing down a table of powers of ten, say from 10^3 to 10^{-3}—including 10^0—and studying the pattern, you can tell what 10^0 has to be. Let's try it.

$$10^3 = 10 \times 10 \times 10 = 1000$$

$$10^2 = 10 \times 10 = 100$$

$$10^1 = 10 = 10$$

$$10^0 = ?$$

$$10^{-1} = \frac{1}{10} = 0.1$$

$$10^{-2} = \frac{1}{10} \times \frac{1}{10} = \frac{1}{100} = 0.01$$

$$10^{-3} = \frac{1}{10} \times \frac{1}{10} \times \frac{1}{10} = \frac{1}{1000} = 0.001$$

Now, search for a pattern. As you move from 10^3 to 10^2 to 10^1 in the table, the whole number changes from 1000 to 100 to 10. This means that each decrease by one in the exponent results in a division of the whole number by ten. So, as 10^3 changes to 10^2 (3 decreasing by 1 to become 2),the 1000 changes to 100—a division by 10. Does the pattern continue as you move from 10^{-1} to 10^{-2} to 10^{-3}?

It does! A decrease of 1 in the exponent leads to a division of the number by 10. For example, as one changes from 10^{-1} to 10^{-2}, the number changes from 0.1 to 0.01, a division by 10.

If you apply the *same pattern* in going from 10^1 to 10^0 (exponent decreases by 1), the number must change from 10 to 1 (division of number by 10). If that's so—and it is—then $10^0 = 1$!

If you let $10^0 = 1$, does the same pattern continue from 10^0 to 10^{-1}? Yes, as you move from 10^0 to 10^{-1} (exponent decreases by 1), the number changes from 1 to 0.1 (division of number by 10).

So, by letting $10^0 = 1$, the same pattern (*decrease of exponent by 1 means division of number by 10)* holds uniformly as you change from any positive power of ten to any negative power of ten.

10^0 is another name for 1.

WRITING NUMBERS IN POWER-OF-TEN NOTATION

Now that we've learned about powers of ten, let's go back to the problem your calculator couldn't handle without scientific notation. Recall that you wanted to calculate the number of atoms in 3.45 ounces of gold, but couldn't enter the very large number 86,700,000,000,000,000,000,000 in your calculator.

To begin to change this large number to scientific notation, let's rewrite the number in this way:

$$86{,}700{,}000{,}000{,}000{,}000{,}000{,}000 =$$

$$867 \times 10 \times 10 \times 10 \times 10 \text{ ...and so forth}$$

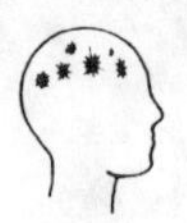

How many zeros do you have to put at the end of 867 to write the number 86,700,000,000,000,000,000,000? How many times do you multiply 867 by 10 to get this number? The next section shows you the answer to this question.

Exponents of ten—large numbers

When you multiply a whole number by ten *once*, how many zeros do you write at the end of the number?

$$867 \times 10 = 8670$$

Try multiplying by ten again.

$$8670 \times 10 = 86700$$

Each time you multiply a whole number by 10, the new whole number has *one* more zero written at the end (at the right).

If you put a decimal point at the end of a whole number (since 867 is the same as 867.), you can see that **multiplying by ten is the same as moving the decimal point one place to the right.** Multiplying by ten makes a number *larger* so the decimal point moves to the *right* when you multiply by ten.

The very large number has 20 zeros after 867. Therefore, 86,700,000,000,000,000,000,000 is the same as 867 multiplied by 10 *twenty* times.

That isn't much of a shortcut! Writing ($\times$ 10) twenty times is no better than writing twenty zeros. What is a shorter way of writing ($\times$10) twenty times? You've already learned that writing down 10 twenty times and multiplying is the same as writing 10^{20}.

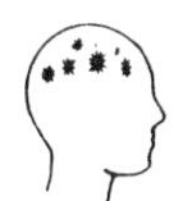

Since 86,700,000,000,000,000,000,000 is the same as 867 multiplied by 10 twenty times, and since multiplying by 10 twenty times can be written as 10^{20}, you can write this:

$$86{,}700{,}000{,}000{,}000{,}000{,}000{,}000 = 867 \times 10^{20}$$

That really is a shortcut! However, the number is not yet written in scientific notation. That's because the multiplier 867 is not a number between 1 and 10.

Soon you'll see how to write numbers in scientific notation. But first let's look at how to write very *small* numbers with exponents of ten.

Exponents of ten—small numbers

Figure 12-3 shows a drawing of red blood cells as they might appear in an electron microscope. The picture shows the cells magnified about 2500 times.

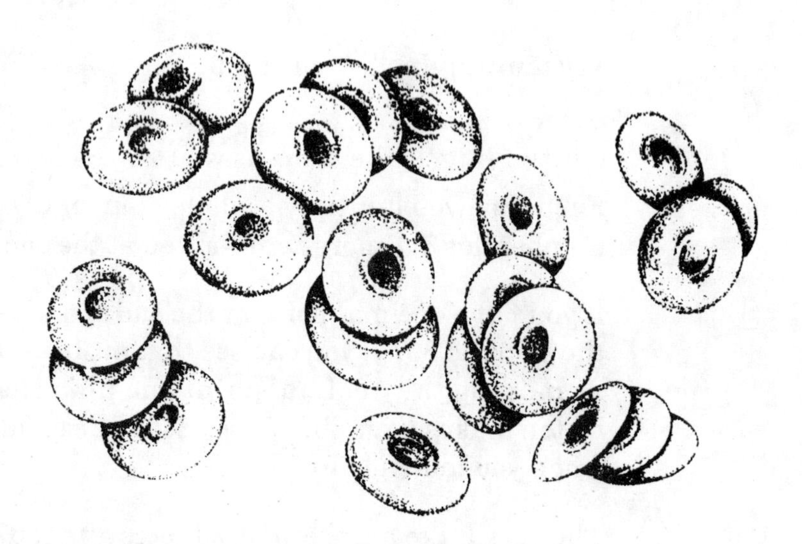

Figure 12-3
Human red blood cells

A single human red blood cell, like one of those pictured in Figure 12-3, is 0.000007 meters in diameter. How can you rewrite this diameter by putting an exponent on ten? You can find the answer by exploring what happens to the decimal point when you *divide* a number, such as 7, by ten.

You can write the number 7 as "7." or "7.0" so you can see the decimal point. What is $7.0 \div 10$ or, saying it another way, what is $7 \times {}^{1}/_{10}$?

$$7 \times \frac{1}{10} = 0.7$$

When you multiply a number by ${}^{1}/_{10}$ once, what happens to the decimal point?

Try multiplying by ${}^{1}/_{10}$ again.

$$0.7 \times \frac{1}{10} = 0.07$$

Now you can see the pattern! **Multiplying a number by ${}^{1}/_{10}$ is the same as moving the decimal point one place to the left.** Multiplying a number by ${}^{1}/_{10}$—or dividing the number by 10—makes a number *smaller* so the decimal point moves to the *left*.

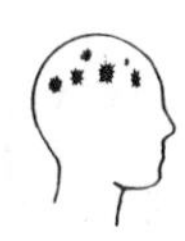

Think again about the number for the diameter of the red blood cell—0.000007 meters. The previous discussion shows that

0.000007 can be written as 7 multiplied by $^1/_{10}$ six times.

$7 \times {}^1/_{10} \times {}^1/_{10} \times {}^1/_{10} \times {}^1/_{10} \times {}^1/_{10} \times {}^1/_{10}$

But you know that writing down the factor $^1/_{10}$ six times, and multiplying, is the same as writing the power of ten equal to 10^{-6}.

So, $7 \times {}^1/_{10} \times {}^1/_{10} \times {}^1/_{10} \times {}^1/_{10} \times {}^1/_{10} \times {}^1/_{10}$ is the same as

$$7 \times 10^{-6}$$

Thus you can write the diameter of a human red blood cell—given as 0.000007 meters—as 7×10^{-6} meters. That's shorter than 0.000007 meters. And, as it happen, it's in scientific notation because the multiplier 7 is between 1 and 10.

WRITING NUMBERS IN SCIENTIFIC NOTATION

So far you've learned what positive powers of ten—such as 10^4 or 10^6—mean, and what negative powers of ten—such as 10^{-3} or 10^{-5}—mean. You've also learned how to write numbers like 86,700,000,000,000,000,000,000 and 0.000007 in power-of-ten notation—as 867×10^{20} and 7×10^{-6}, respectively. The first of these is not yet in scientific notation, the second one is. Let's now look closely at numbers in *scientific notation.*

Numbers in scientific notation

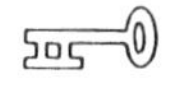

A number is written in scientific notation when it is written in this way: **a number between one and ten multiplied by a power of ten.**

Look again at the number 867×10^{20}. Is this number written in scientific notation? The number 867 is multiplied by a power of ten, but 867 is not a number between one and ten. This means that 867×10^{20} is not written in scientific notation.

To write this number in scientific notation, you must begin with a number that is between one and ten. How can you write 867×10^{20} in a form that begins with a number between one and ten?

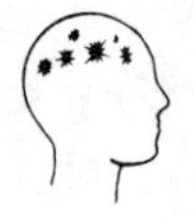

To write 867 in a different way, you can use the fact that **multiplying a number by ten is the same as moving the decimal point one place to the right.** This fact suggests that you can rewrite 867 in this way: move the decimal point to the *left* to make the number *smaller* and, at the same time, multiply the number by ten to make the number *larger.* In this way you can rewrite a number without changing its value.

Thus

$$86.7 \times 10 = 867$$

But 86.7 is still not a number between one and ten.

Move the decimal to the left one more place to make the number still smaller, and multiply one more time by ten to keep the value the same. This gives

$$867 = 8.67 \times 10 \times 10$$

$$\text{or } 8.67 \times 10^2$$

Since 8.67 is a number between one and ten, you are now ready to write the original number (86,700,000,000,000,000,000,000) in scientific notation. Putting it all together:

$$86{,}700{,}000{,}000{,}000{,}000{,}000{,}000 = 867 \times 10^{20} = 8.67 \times 10^2 \times 10^{20}$$

Multiplying by 10 twice, and then by 10 twenty times as in $10^2 \times 10^{20}$, is the same as using 10 as a factor 22 times. This means that $10^2 \times 10^{20} = 10^{22}$.

A shorter way to write this number is

$$8.67 \times 10^{22}$$

Is 8.67×10^{22} written in scientific notation? Is it written as a **number between one and ten multiplied by a power of ten?** Since 8.67 is a number between one and ten, and 10^{22} is a power of ten, 8.67×10^{22} *is* written in scientific notation.

How to write in scientific notation

Now let's look at a *general method* for writing numbers in scientific notation. Example 3 shows how to do this for small numbers. Example 4 shows how to do the same thing for large numbers.

Example 3:
Writing a small number in scientific notation

Let's examine the process by rewriting the number 0.000007 in scientific notation. Follow these steps:

First, write the number with a pointer ($_\wedge$) just to the *right of the first digit that is not zero*. This will be the new location of the decimal point.

0.000007$_\wedge$ meters

Second, use your pencil point to move from the pointer (the new decimal point) to the old decimal point, counting how many places you move, and in which direction (left or right).

0.000007$_\wedge$ meters

Notice that the new decimal point has made the number *larger* than the old number.

How many places did you move as you went from the new position (at the $\wedge$) to the old position of the decimal point? By moving the decimal you have made the number *larger* by six places. To balance this, you must make the number *smaller* by *dividing* it by ten six times.

Dividing by ten six times is the same as multiplying by $^1/_{10}$ six times. And multiplying by $^1/_{10}$ six times is the same as multiplying by 10^{-6} since

$$\frac{1}{10} \times \frac{1}{10} \times \frac{1}{10} \times \frac{1}{10} \times \frac{1}{10} \times \frac{1}{10} = 10^{-6}$$

So, you can replace 0.000007$_\wedge$ meters by

$7_\wedge \times 10^{-6}$ meters

Third, replace the pointer with the new decimal point. The final result is:

7×10^{-6} meters

Finally, check to see if the new number is in scientific notation—a number between one and ten multiplied by a power of ten.

Here is a summary of those steps:

To convert from decimal form to scientific notation:

First, put a pointer ($_\wedge$) to the *right of the first nonzero digit.*

Second, count the spaces as you move from the pointer ($_\wedge$) to the old decimal point.

Third, replace the pointer ($_\wedge$) with the new decimal point and multiply the number by 10 with an exponent that is the same as the number of places you moved. Moving *to the left* from the pointer (the new decimal point) gives a negative exponent; moving *to the right* from the pointer gives a *positive exponent.*

Fourth, check to see if the new number is in scientific notation—a number between one and ten multiplied by a power of ten.

Example 4: ***Rewriting a large number in scientific notation***

Try using the preceding steps to rewrite 270,000,000 molecules (the number of hemoglobin molecules in a single human red blood cell) in scientific notation.

First, put a pointer to the *right of the first nonzero digit.*

$$2_\wedge 70{,}000{,}000 \text{ molecules}$$

Second, count the spaces as you move from the pointer ($_\wedge$) to the old decimal point. (In this example you have to write in the *understood* decimal point—just to the right of the last zero.) As you move from the pointer ($_\wedge$) to the decimal point, you count 8 spaces.

$$2_\wedge 70{,}000{,}000. \text{ molecules}$$

Third, replace the pointer with the new decimal point and multiply the number by 10 with an exponent that is the same as the number of places you moved—in this case, 8. Remember, moving to the left gives a negative exponent; moving to the right gives a positive exponent. The final answer is

$$2.7 \times 10^8 \text{ molecules}$$

Fourth, check to see if the new number is in scientific notation—a number between one and ten times a power of ten.

How to rewrite in decimal form

Sometimes it is useful to rewrite numbers in the opposite direction, that is, from scientific notation to decimal form.

To convert from scientific notation to decimal form:

First, put a pointer in place of the original decimal point.

Second, move your pencil left or right from the pointer (left if the exponent is negative, right if the exponent is positive) as many places (adding zeros as needed) as the *absolute value* of the exponent.

Third, put a new decimal point where you stop moving.

Fourth, check to make sure that the new number is *equal* to the original number written in scientific notation.

Try applying these steps to rewrite this number in decimal form:

$$1.6022 \times 10^{-19} \text{ joules}$$

Example 5: ***Changing from scientific notation to a decimal number***

First, put a pointer in place of the original decimal point.

$$1_{\wedge}6022 \times 10^{-19} \text{ joules}$$

Second, move left from the pointer (since the exponent is negative) 19 places (19 is the absolute value of –19), adding zeros as needed.

$$0000000000000000001_{\wedge}6022 \text{ joules}$$

Third, put a new decimal point to the left of the last zero you wrote and add a zero in front of the decimal point to protect it from being accidentally omitted.

$$0.00000000000000000016022 \text{ joules}$$

Fourth, check to make sure that the number you write is *equal* to the original number written in scientific notation.

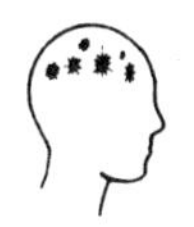

Whether you are rewriting numbers from decimal form to scientific notation, or from scientific notation to decimal form, remember these things:

- The pointer ($_{\wedge}$) goes just to the *right of the first nonzero digit.* If you are changing to scientific notation, this will be the new decimal point. If you are changing to the decimal form, the pointer will be at the original decimal point.

- You move from the pointer to the decimal point. (You move from the pointer to the old decimal point if you are changing to scientific notation; you move from the pointer to the new decimal point if you are changing to the decimal form.)
- A move to the right corresponds to a positive exponent; a move to the left corresponds to a negative exponent.
- You move as many places as the *absolute value* of the exponent.

Study Activity: Complete the following exercises to check your understanding of writing numbers in scientific notation—and changing from numbers in scientific notation to decimal form.

1. Write the following numbers in scientific notation.

   a. 56.7

   b. 56.7×10^2

   c. 86,700

   d. 0.00567

   e. 10.1×10^{-6}

   f. 0.00236×10^{-6}

2. Change the following numbers written in scientific notation to decimal form.

   a. 5.67×10^1

   b. 7.2×10^5

   c. 1.23×10^{-4}

   d. 8.67×10^{22}

   e. 9.1×10^{-12}

   f. 1.02×10^{-1}

USING SCIENTIFIC NOTATION WITH YOUR CALCULATOR

Now that you can write numbers in scientific notation, you can use your calculator to work problems that have numbers with many zeros.

Entering numbers in scientific notation

Study Activity:

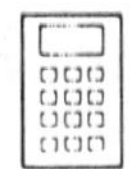

Follow these steps to enter 1.6022×10^{-19} joules in your calculator.

First, enter the first number (the one between one and ten).

Enter 1.6022

Second, tell the calculator that you are about to enter an exponent by pressing the key labeled EE (which stands for Enter Exponent).

Press the EE key

Third, enter the exponent of ten—in this case, the number 19.

Enter 19
Press the +/– key to make the exponent negative.

Look at the display of your calculator to see how the calculator shows the number 1.6022×10^{-19} in the window.

To enter a number in scientific notation in the calculator:

First, enter the first number (the one between one and ten).

Second, press the key labeled EE.

Third, enter the exponent of ten and, if necessary, press the +/– key to make the exponent negative.

Limitations on working problems with the calculator

Now you are going to enter some numbers to see what your calculator can handle.

Study Activity:

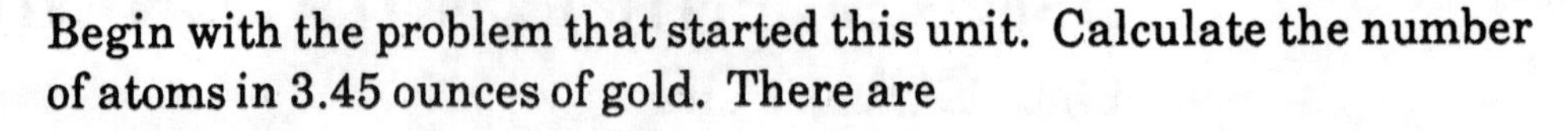

Begin with the problem that started this unit. Calculate the number of atoms in 3.45 ounces of gold. There are

86,700,000,000,000,000,000,000 atoms in one ounce of gold. Begin by converting 86,700,000,000,000,000,000,000 atoms to scientific notation.

8$_\wedge$6,700,000,000,000,000,000,000. atoms per ounce

is the same as 8.67×10^{22} atoms per ounce

Then, enter this number into your calculator.

Enter 8.67
Press the EE key
Enter 22

Now multiply this number by 3.45 ounces.

Press the $\times$ key
Enter 3.45
Press the $=$ key

Read the answer in the display window and write it as a number in scientific notation.

2.99115×10^{23} atoms

You can enter a number using the EE (Enter Exponent) key even if the number is not in scientific notation. However, many calculators will convert the number to scientific notation as soon as you start to perform an operation (such as $=$ or $\times$ or $\div$). The next activity shows this.

Study Activity:

Enter 2456×10^{80} in your calculator.

Enter 2456
Press the EE key
Enter 80

The display window shows *2456 80* which is not in scientific notation since 2456 is not a number between one and ten.

Press the $=$ key

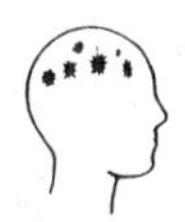

What has changed in the display window? With many calculators, the display window changes to show *2.456   83* because the calculator rewrites the number in scientific notation as soon as you press a key for any operation. Now try multiplying that number by 2.

Press the × key
Enter 2
Press the = key

The calculator shows *4.912   83* in the window.

What are the limits on the exponent (83 in this case)? Let's experiment to find out.

Multiply 4.912×10^{83} by 1.0×10^{16} to see what happens. With *4.912   83* in the window,

Press the × key
Enter 1
Press the EE key
Enter 16
Press the = key

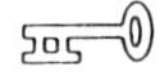

The calculator displays *4.912   99* in the window. Many calculators use only the two spaces at the far right of the window for the exponent. This means that 99 is the largest exponent that can be displayed.

To see what happens when the exponent is larger than 99, try multiplying this number 4.912×10^{99} by 1×10^{1} with these additional steps:

Press the × key
Enter 1
Press the EE key
Enter 1
Press the = key

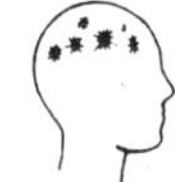

What happens? The display window shows an E (that stands for **Error Signal**). You asked the calculator to work a problem that was too big for it to handle.

To clear this error condition,

Press the *All Clear* key

If the exponent of ten is **more than 99 or less than –99**, most calculators cannot handle the number in scientific notation.

Of course, the first part of a number in scientific notation (if it is a positive one) must be **equal to or less than 9.9999** if there are 5 places saved in the display window for this part of a number.

Most calculators can also handle a **negative number** written in scientific notation. The negative number between one and ten must not be **smaller than –9.9999** if there are 5 places saved in the display window for this part of a number.

COMBINING NUMBERS IN SCIENTIFIC NOTATION

When you write a number in scientific notation, you are combining numbers by multiplying them. The number between one and ten is *multiplied* by the power of 10 shown by the exponent.

You have also seen that dividing by 10 can be written as multiplying by 10^{-1} when a number is written in scientific notation.

This means that you can multiply and divide numbers written in scientific notation with your calculator without learning any new rules.

Multiplying numbers in scientific notation

You have already used the fact that $10^2 \times 10^{20}$ is 10^{22} because those exponents are shortcuts for writing ten two times and then twenty times and then multiplying all those tens (22 of them). This meaning of exponents leads to a rule for multiplying two numbers that have the same base but that may have *different exponents*.

To multiply two powers of 10, add the exponents.

Study Activity: Try this out for yourself. What is $10^2 \times 10^3$?

The meaning of the exponent 2 in 10^2 tells you that

$$10^2 = 10 \times 10$$

and the meaning of the exponent 3 in 10^3 tells you that

$$10^3 = 10 \times 10 \times 10,$$

so it must be true that

$$10^2 \times 10^3 \text{ is the same as}$$

$$(10 \times 10) \times (10 \times 10 \times 10)$$

which, by the definition of exponent, is the same as

$$10^5$$

Written the short way, the problem looks like this:

$$10^2 \times 10^3 = 10^{2+3} = 10^5$$

To multiply two powers of 10, add the exponents.

Now let's use that rule to multiply two numbers written in scientific notation.

Study Activity:

There are 2.7×10^8 hemoglobin molecules in a single human red blood cell, and there are about 5.0×10^6 human red blood cells in one cubic millimeter of blood. How many hemoglobin molecules are in one cubic millimeter of blood?

To work this problem, multiply

$$2.7 \times 10^8 \text{ molecules} \times 5.0 \times 10^6 \text{ cells in one mm}^3$$

to find the molecules in one cubic millimeter of blood.

Rewrite this problem to group the numbers between one and ten together and the powers of ten together.

$$(2.7 \times 5.0) \times (10^8 \times 10^6) \text{ molecules}$$

Use your calculator to multiply the numbers in the first parentheses. You can either multiply $10^8 \times 10^6$ in your head by adding the 8 and 6 to get 10^{14}, *or* you can use your calculator.

To enter 10^8 in the calculator, you must enter it as 1×10^8. The calculator needs a 1 (or some other number) each time before you press the EE key.

Did you get an answer of 13.5×10^{14} molecules?

You can work a problem with negative exponents in exactly the same way.

Study Activity: Multiply 3.567×10^{-4} by 8.5×10^{-3}

Begin by rewriting the problem to group the powers of ten together.

$$(3.567 \times 8.5) \times (10^{-4} \times 10^{-3})$$

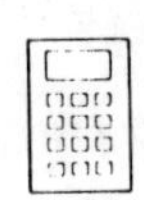

Then use your calculator as needed to multiply within the two sets of parentheses.

$$(30.3195) \times (10^{-4+(-3)})$$

$$30.3195 \times 10^{-7}$$

This answer is *not* in scientific notation because the first number (30.3195) is more than one. Rewrite this number in scientific notation

$$3.03195 \times 10^1 \times 10^{-7}$$

and then combine the powers of ten by adding the exponents.

$$3.03195 \times 10^{1+(-7)}$$

$$3.03195 \times 10^{-6}$$

Now let's work this same problem again, but this time let the calculator do the work!

Multiply 3.567×10^{-4} by 8.5×10^{-3}

Enter 3.567
Press the EE key
Enter 4
Press the +/– key
Press the × key
Enter 8.5
Press the EE key
Enter 3
Press the +/– key
Press the = key

The display shows *3.032 –06* which is the same answer as before except that 3.03195 is rounded to 3.032 and the exponent of –6 is written as –06 in the window.

Dividing numbers in scientific notation

How do you divide one power of ten by another? Try an example to see what happens.

Study Activity:

What is $10^5 \div 10^3$?

The meaning of the exponent tells you that

$$10^5 = 10 \times 10 \times 10 \times 10 \times 10$$

and also that

$$10^3 = 10 \times 10 \times 10$$

so it must be true that

$$10^5 \div 10^3 \text{ is the same as}$$

$$\frac{10 \times 10 \times 10 \times 10 \times 10}{10 \times 10 \times 10}$$

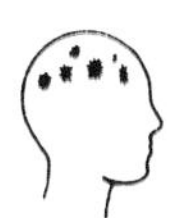

This fraction can be reduced by using the fact that $^{10}/_{10} = 1$ and rewritten this way:

$$1 \times 1 \times 1 \times 10 \times 10$$

$$\text{or } 10 \times 10 = 10^2$$

Putting this together,

$$10^5 \div 10^3 = 10^2$$

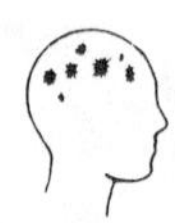

What operation between the exponents 5 and 3 would give the answer of 2? What must replace the "?" in this sentence: 5 ? 3 = 2.

$$10^5 \div 10^3 = 10^{5-3} = 10^2$$

See if this same method works on another problem.

What is $10^4 \div 10^7$?

Rewrite this problem as a fraction—

$$\frac{10 \times 10 \times 10 \times 10}{10 \times 10 \times 10 \times 10 \times 10 \times 10 \times 10}$$

Simplify this fraction by using the fact that $^{10}/_{10} = 1$. You should get

$$\frac{1}{10 \times 10 \times 10}$$

This is the same as $^1/_{10^3}$, which you know (from the table you wrote) can be rewritten as 10^{-3}.

This means that

$$10^4 \div 10^7 = 10^{4-7} = 10^{-3}$$

Both these examples suggest this rule:

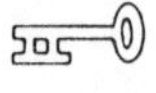

To divide one power of 10 by another power of 10, subtract the exponent of the second (the divisor) from the exponent of the first (the dividend).

This rule is easier to write in symbols:

$$10^a \div 10^b = 10^{a-b}$$

Here is the rule for multiplying powers, written in symbols:

$$10^a \times 10^b = 10^{a+b}$$

Let's apply the rule for dividing powers of ten to a problem that has numbers written in scientific notation.

Study Activity:

If you have a cube of gold that has a mass of 2.41 grams (the cube measures 5 mm on an edge), and there are 7.4×10^{21} atoms of gold in the cube, what is the mass of one atom of gold?

$$\text{The mass of one atom} = \frac{\text{mass of the cube}}{\text{number of atoms in the cube}}$$

Begin by changing 2.41 into scientific notation. $2.41 = 2.41 \times 10^0$

$$\frac{2.41 \times 10^0 \text{ grams}}{7.4 \times 10^{21} \text{ atoms}}$$

Use your calculator to find $2.41 \div 7.4$ and subtract the exponents.

$$(2.41 \div 7.4) \times (10^0 \div 10^{21}) \text{ grams per atom}$$

$$0.32568 \times 10^{(0-21)} \text{ grams per atom}$$

$$(3.2568 \times 10^{-1}) \times 10^{-21} \text{ grams per atom}$$

$$3.2568 \times (10^{-1} \times 10^{-21}) \text{ grams per atom}$$

$$3.2568 \times 10^{-1+(-21)} \text{ grams per atom}$$

$$3.2568 \times 10^{-22} \text{ grams per atom}$$

Now let's work this same problem again, but this time let the calculator do the work!

Divide 2.41×100 grams by 7.4×10^{21} atoms

Enter 2.41
Press the EE key
Enter 0
Press the ÷ key
Enter 7.4
Press the EE key
Enter 21
Press the = key

The display shows *3.2568 –22* which is the same answer as before:

$$3.2568 \times 10^{-22} \text{ grams per atom}$$

CONVERTING METRIC MEASUREMENTS

Exponents of ten and scientific notation make it easier to convert measurements within the metric system. Figure 12-4 reminds you of the various names in the metric system.

VALUE	EXPONENT	SYMBOL	PREFIX
1 000 000 000 000	10^{12}	T	tera
1 000 000 000	10^{9}	G	giga
1 000 000	10^{6}	M	mega
1 000	10^{3}	k	kilo
100	10^{2}	h	hecto
10	10^{1}	da	deca
0.1	10^{-1}	d	deci
0.01	10^{-2}	c	centi
0.001	10^{-3}	m	milli
0.000 001	10^{-6}	μ	micro
0.000 000 001	10^{-9}	n	nano
0.000 000 000 001	10^{-12}	p	pico

Figure 12-4
Names in the metric system

These names can be written before any metric unit of measure, such as the meter, the gram, or the liter. They can also be used for units of time—such as seconds—and for units used to measure electricity—such as ohms, amperes, volts, and watts.

Ohms and amperes are used in electronics. A megohm is a million ohms. A microamp is one one-millionth of an ampere. In computers, the data moves at speeds measured in nanoseconds (1×10^{-9} seconds) and even picoseconds (1×10^{-12} seconds).

The chart in Figure 12-5 can help you convert from one metric unit to another.

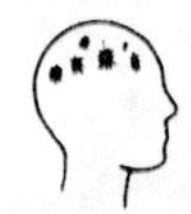

You can use the information in Figure 12-5 to extend this conversion chart to include micro, nano, and pico if you need those units.

Here are the steps to use the chart in Figure 12-5:

First, find the name in the column at the left for the unit you wish to convert.

Second, go across the chart until the top of the chart has the name of the unit you need.

TO CONVERT	TO MILLI-	TO CENTI-	TO DECI-	TO METER GRAM LITER	TO DECA-	TO HECTO-	TO KILO-
KILO-	x 10^6	x 10^5	x 10^4	x 10^3	x 10^2	x 10^1	
HECTO-	x 10^5	x 10^4	x 10^3	x 10^2	x 10^1		x 10^{-1}
DECA-	x 10^4	x 10^3	x 10^2	x 10^1		x 10^{-1}	x 10^{-2}
METER GRAM LITER	x 10^3	x 10^2	x 10^1		x 10^{-1}	x 10^{-2}	x 10^{-3}
DECI-	x 10^2	x 10^1		x 10^{-1}	x 10^{-2}	x 10^{-3}	x 10^{-4}
CENTI-	x 10^1		x 10^{-1}	x 10^{-2}	x 10^{-3}	x 10^{-4}	x 10^{-5}
MILLI-		x 10^{-1}	x 10^{-2}	x 10^{-3}	x 10^{-4}	x 10^{-5}	x 10^{-6}

Figure 12-5
Conversion chart for metric units

Third, multiply the unit you wish to convert by the power of ten shown in the box.

Study Activity:

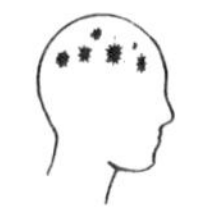

Change 35 milliamps to amperes.

First, find milli in the column at the left.

Second, look across to try to find a box that is labeled *ampere.* An ampere is another basic metric unit like a meter, gram, or liter, so you can add ampere (or any other metric unit) to the center box on this chart.

Third, multiply as shown in the box.

$$35 \text{ milliamps} \times 10^{-3} = 35 \times 10^{-3} \text{ amperes}$$

You can rewrite this as

$$3.5 \times 10^1 \times 10^{-3} \text{ amperes}$$

which is the same as

$$3.5 \times 10^{-2} \text{ amperes or } 0.035 \text{ amperes}$$

SUMMARY

Scientific notation is a short, efficient way to write and calculate with very large and very small numbers. A number is written in scientific notation as a number between 1 and 10, multiplied by some power of 10.

Numbers expressed in scientific notation that have positive exponents—for example 2.98×10^{13}—are large numbers. Numbers expressed in scientific notation that have negative exponents—for example, 1.6×10^{-19}—are small numbers.

The exponent gives the order of magnitude or size of the number. The number 3.25×10^{3} means that there are 3.25 *thousands*. The number 2.8×10^{-2} means that there are 2.8 *hundredths*.

Since *place value* is based on powers of ten, the decimal point can be moved to the right or left to express numbers between one and ten. The number is then multiplied by the power of ten that leaves the original number's value unchanged.

Whether you are rewriting numbers from decimal form to scientific notation, or from scientific notation to decimal form, remember these facts:

- A move to the right corresponds to a positive exponent; a move to the left corresponds to a negative exponent.
- You move as many places as the absolute value of the exponent.

Numbers can be entered into your calculator using the EE key. Products and quotients can be calculated by using the × or ÷ key between entries. The +/– key allows you to enter both positive and negative exponents and to find their products and quotients.

Metric conversions from one unit to another are easily made from a table or conversion chart. Examples are 1 km = 10^{3} meters or 1 ml = 10^{-3} liters. An expanded table shows that exponents range from 10^{-12} (prefix-pico) to 10^{12} (prefix-tera). This represents 24 orders of magnitude between the smallest and largest units.

PRACTICING THE SKILLS

Laboratory Activities

Use the mathematics skills you have learned to complete one or all of the following activities:

Activity 1: Measuring average paper thickness

Equipment

Vernier calipers (calibrated in inches)
Micrometer calipers (calibrated in inches)
Chalkboard
Chalk
Four books, catalogs, or phone books—labeled A, B, C and D—each with 500 or more numbered pages.
Calculator

Statement of Problem

In this activity, you find the average paper thickness in each of four books. You do this by measuring the thickness of 500 numbered pages, then finding the average paper thickness. You compare your answer to the paper thickness measured with a micrometer caliper.

Procedure

a. Use the vernier calipers to measure the total thickness of numbered pages 1-500 for each book. Convert the measurement in fractions of an inch to a decimal number rounded to the nearest thousandth of an inch. Write the decimal number thickness for each book on the chalkboard under the correct heading (Book A, Book B, Book C, or Book D).

b. Calculate an average paper thickness for each book by dividing the measured thickness by one-half the numbered pages measured—remember there are two numbered pages for each sheet of paper. Some may be printed one-sided. Write your group's answer for the average paper thickness, rounded to the nearest thousandth of an inch—for each book—in **scientific notation**. List it on the chalkboard.

c. After all groups in your class have written the average paper thickness for each book on the chalkboard, calculate an overall **class** average of the paper thickness for each book. Round to the nearest thousandth of an inch. Write the class-averaged paper thickness for each in **scientific notation**. List values on the chalkboard.

d. How does the average paper thickness *calculated by your group* compare to the *class-averaged paper thickness* for each book? Which value do you think should be more correct?

e. Use micrometer calipers—that probably measure to a thousandth of a inch—to measure the thickness of a single sheet of paper in each book. (See your teacher if you have trouble reading a micrometer caliper.) Convert your answer in fractions of an inch to a number in **scientific notation**. How does the paper thickness measured directly with the micrometer caliper compare with your group's calculated average paper thickness? Which do you think is more correct?

Activity 2: Counting sand grains

Equipment

Graduated cylinder, 10-milliliter capacity
Coarse sand (such as cat litter)
Calculator
Micrometer calipers

Statement of Problem

In this activity, you count the number of grains of sand in a small volume and use this number to **estimate** the number of grains in a larger volume. You also estimate the volume of an average grain of sand.

Procedure

a. Use the graduated cylinder to measure 1 milliliter of sand. Count the grains of sand in the measured 1 milliliter.

b. Based on your count, how many grains of sand would you estimate are in a cubic meter? (Remember, 1 milliliter is equal to 1 cubic centimeter.) State your answer in **scientific notation**.

c. Based on your count, how many grains of sand would you estimate are in a strip of beach 1500 meters by 200 meters down to a depth of 10 centimeters? Give your answer in **scientific notation**.

d. Based on your answer to **a**, what is the average volume (in cubic centimeters) of a grain of sand? State your answer in **scientific notation**.

e. Isolate 4 or 5 typical grains of sand and measure their diameter with a micrometer caliper. (See your teacher if you need help in using a micrometer caliper.) From these measurements, calculate an average diameter. Then use the formula for the volume of a sphere ($V = \frac{1}{6} \pi d^3$) to get an approximate volume for an average grain of sand.

Write your answer—in cubic centimeters—in **scientific notation.**

f. How do the values for Parts **d** and **e** compare?

Activity 3: Measuring water molecules

Equipment

Graduated cylinder, 500-milliliter capacity
String
Spring scales with a 5000-gram capacity
Water supply and drain
Calculator

Statement of Problem

In this activity, you measure the volume and weight of a sample of water and use this data to calculate the **number of molecules** in the sample and the **average volume and weight of each molecule.**

Procedure

a. Tie a string around the top of the graduated cylinder as shown below. Weigh the empty graduated cylinder and string by hooking the loop on the spring scales. Write this weight on a sheet of data paper.

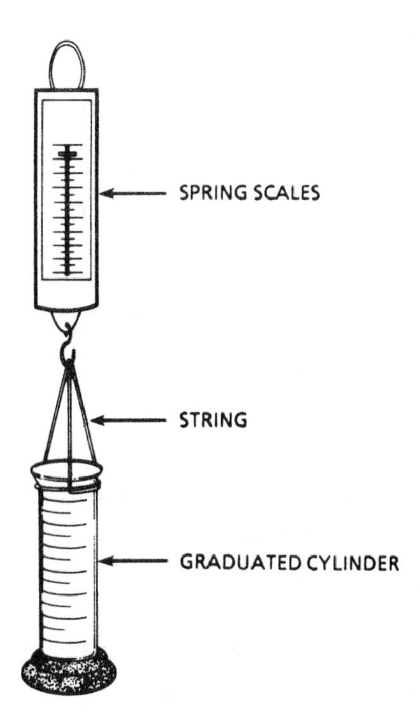

b. Measure out a 500-milliliter volume of water in the graduated cylinder. Weigh the water and graduated cylinder together on the spring scales. Write this combined weight on a sheet of data paper.

c. Subtract the weight of the empty graduated cylinder and string from the combined weight of the water and graduated cylinder. This is the weight of the water sample. There are 6.02×10^{23} molecules of water in 18 grams of water. Set up a proportion to determine the number of molecules in your 500-milliliter sample. Write your answer in **scientific notation.**

d. Based on the data in Step c, calculate the average volume per molecule, and the average weight per molecule. Write your answer in **scientific notation.**

Student Exercises

You can solve the exercises that follow by applying the mathematics skills you've learned. The problems described here are those you may meet in the world of work.

NOTE: *Wherever possible, use your calculator to solve the problems that require numerical answers.*

GENERAL

EXERCISE 1: The length of the tropical year is 365.24220 days, as compared to the length of 365.2425 days used by the Gregorian calender.

a. What is the difference between the tropical year and the Gregorian calender year? (Express your answer in scientific notation.)

b. How many years will it take for this small difference to amount to one day?

EXERCISE 2: Some home heaters have been found to produce pollutants into the room air. A maximum acceptable level of carbon monoxide, for example, has been suggested to be about 9 parts per million. For nitrogen dioxide a value of about 0.05 parts per million has been suggested. Express each of these concentrations as a percentage, in scientific notation.

EXERCISE 3: A laser printer prints characters at a density of 300 dots per inch.

a. How many dots can be printed across the width of 8.5 inches of a standard letter-sized piece of paper?

b. How many dots can be printed down the length of 11.0 inches?

c. If every dot was printed on the paper, how many dots would be printed? (Express your answer in scientific notation.)

d. If each dot just touched the adjacent dot, what would be the diameter of a dot? (Express your answer in scientific notation.)

EXERCISE 4: A phonograph needle is essentially a smooth round surface. A particular needle is 0.02 cm wide. A measurement of the tracking force on a phonograph record yields 15 grams.

a. The needle tip is essentially half of a sphere. Convert the diameter of the half-sphere to inches. What is the surface area of the tip of the needle, in square inches? (Express your answer in scientific notation.)

b. Convert the tracking force to pounds of force. (Express your answer in scientific notation.)

c. Divide the tracking force by the area of the needle tip to find the pressure exerted by the needle on the phonograph record (in pounds per square inch, or psi).

EXERCISE 5: A trip to the nearest star has been proposed. The nearest star is about 4 $^{1}/_{2}$ light years away. A light year is the distance that light can travel in a year. The speed of light is approximately 186,000 miles per second.

a. Convert the speed of light to miles per year by multiplying the speed per second by the number of seconds in a year. This is how far light travels in a year. (**Hint:** Consider the number of days in a year, the number of hours in a day, etc.) (Express your answer in scientific notation.)

b. How far away (in miles) is the nearest star? (Express your answer in scientific notation.)

c. The spaceships that visited the moon traveled nearly 30,000 miles per hour. How many hours would it take at this speed to reach this nearest star?

AGRICULTURE AND AGRIBUSINESS

EXERCISE 6: A recent report of the leading corn-growing states showed the following annual amounts grown for grain:

State	Annual growth (bushels)
Iowa	1,739,900,000
Illinois	1,452,540,000
Nebraska	802,700,000
Minnesota	744,700,000
Indiana	654,000,000
Wisconsin	378,000,000
Ohio	360,000,000
Michigan	273,600,000
Missouri	213,400,000
South Dakota	180,000,000

a. What is the total corn produced by these states? (Express your answer in scientific notation.)

b. If each bushel is about 56 pounds, how many pounds of corn did these states produce?

c. The annual production of the entire country was reported as being 8,201,000,000 bushels. How many bushels were produced by states not shown in the above list? (Express your answer in scientific notation.)

EXERCISE 7: Diesel fuel can provide about 138,400 Btu per gallon. A tractor used 215 gallons of diesel fuel over a period of several days, a total of 56 hours running time.

a. How many Btu of energy were provided by the fuel? (Express your answer in scientific notation, rounded to two decimal places.)

b. If 2546 Btu in one hour is equivalent to 1 horsepower, how much horsepower was provided by the fuel during the 56 hours of running time?

c. If the tractor is actually obtaining about 70 horsepower, what is the percentage of horsepower being obtained, compared to the horsepower being provided by the diesel fuel? (This is the thermal efficiency of the engine.)

EXERCISE 8:

You farm about 175 acres in an area that needs irrigation. The typical monthly rainfall for the spring is normally about 7 inches. This is equivalent to water at a depth of 7 inches over your entire 175 acres.

a. How many square feet does your land cover?

b. Convert this to square inches, and multiply by the 7-inch depth of rainfall to obtain the cubic inches of rainfall on your acreage. (Express your answer in scientific notation.)

c. If you had to irrigate this amount of water, how many gallons of water would you have to use?

EXERCISE 9:

A protected coastal wildlife area is surveyed and sampled for various forms of wildlife. The protected area is essentially rectangular in shape, 92 miles long and 3.5 miles wide. Several 50-foot square sections were sampled throughout the area. The average population of a certain insect species in each section was 17.

a. Determine the area of the protected area and convert it to square feet. (Express your answer in scientific notation.)

b. Compute the area of each section that was sampled, and determine the average population of this insect per square foot. (Express your answer in scientific notation.)

c. What is the estimated population of this insect within the boundaries of the protected wildlife area? (Express your answer in scientific notation.)

EXERCISE 10:

In one region of Hong Kong, China, there are 55,000 persons living on an area of about 24 acres.

a. Convert the 24 acres to square feet. (Express your answer in scientific notation.)

b. On the average, how many people are there in each square foot of space in this region?

c. Multiply the answer to Part b by the number of square miles per square foot to determine about how many people are living in each square mile in this region.

d. Repeat the above calculations using the population for your city or town, and the size of your town. How does this compare to the area where you live?

BUSINESS AND MARKETING

EXERCISE 11: You are making a bid to operate the concession stands at a play-off game. You expect a very large crowd, possibly as high as 85,000 fans. Past experience has shown that each fan in this region will purchase an average of about 1.75 cups of soft drink.

a. How many cups of soft drink can you expect to sell? (Express your answer in scientific notation.)

b. If each cup is filled with 6 oz of ice, how many ounces of ice can you expect to use? (Express your answer in scientific notation.)

EXERCISE 12: A newspaper company uses a process that consumes 0.17 pounds of ink for every 1000 pages of print. The newspaper produces about 550 pages per week, with a distribution that averages about 130,000 per day.

a. What is the amount of ink on each page of newsprint (on the average)? (Express your answer in scientific notation.)

b. What is the approximate total number of pages of newsprint produced during a week? (Express your answer in scientific notation.)

c. About how much ink should be used during a week's printing? (Express your answer in both scientific notation and standard notation.)

EXERCISE 13: A report showed that nationwide consumer spending for professional services rose from $260,000,000,000 to $410,000,000,000 over a 10-year period.

a. Express these amounts in scientific notation.

b. What was the average growth in spending on services each year? (Express your answer in scientific notation.)

EXERCISE 14: A newspaper reports that the average price of a new home, across the nation, is $93,000. The report estimates that 65,000 new homes were sold nationwide during a recent period.

a. Approximately how much money was invested in new homes during this period? (Express your answer in scientific notation.)

b. A friend quips that "megabucks" are being spent on new homes. Using your knowledge of prefixes and powers of ten, is your friend right, using the term "megabucks"? If so, how many "megabucks" is it?

EXERCISE 15: In the past several years, the subcompact division of Luxury Motors produced a total of 17,400,000 cars from its 3 assembly plants. Reports of dissatisfaction over the steering wheel design of these models have been reported from 281 owners. What percent of the owners are dissatisfied? (Express your answer in scientific notation.)

EXERCISE 16: A portion of a computer printout has been "formatted" to display its values in scientific notation, as shown below, to use less space on the printout.

```
Beginning inventory    .   7.125 E+04
Additional purchases   .   2.450 E+03
Production adds    . . .   1.035 E+05
```

a. What is the total of the three figures reported above?

b. The data apparently has been rounded off somewhat. Write each of the three numbers in standard notation and indicate where you think the rounding occurred in each number.

c. Suppose that someone wanted to see more digits, that is, less rounding. They suggest reporting the inventory in terms of "thousand units." Rewrite the data shown above, in terms of "thousand units."

EXERCISE 17: Suppose the national budget is $1.8 trillion ($1,800,000,000,000).

a. Express this amount in scientific notation.

b. If the population of the country is 250,000,000, what is each member of the population's portion of the budget?

EXERCISE 18: A report states that credit card users charged a total of $150 billion during a year. During the reported time period, lenders collected a total of about $12.6 billion in interest on outstanding charges.

a. Express each of the dollar figures above in scientific notation.

b. What percent of the charged amount is the interest amount collected by the lenders?

EXERCISE 19: A computer sales brochure advertises that its system has a 16-megabyte memory and can store a total of nearly a half-gigabyte of data on its three hard disk drives.

a. If a single character occupies a byte of memory or disk space, how many characters can be stored in the 16-megabyte memory? (Express your answer in scientific notation and standard notation.)

b. About how many characters can be stored on each of the three identical disk drives? (Express your answer in scientific notation and standard notation.)

EXERCISE 20: A warehouse has 7 modules, each one with 8 rows of 120 bays. Each bay holds 25 pallets. Each pallet contains 40 tires, each tire averaging 35 pounds. Approximately how many pounds of tires can the warehouse hold? (Express your answer in scientific notation.)

HEALTH OCCUPATIONS

EXERCISE 21: A synthetic thyroid preparation that is available in 25-microgram tablets is prescribed to a patient. The bottle of medication contains 30 tablets.

a. Express the amount of medication in each tablet in grams, in scientific notation.

b. How many grams of medication are in each bottle of medication? (Express your answer in scientific notation.)

EXERCISE 22: A group of researchers conducted independent studies of a bacterium. The researchers reported the average length of the bacteria to be:

Researcher	Average reported length
A	0.001633 micrometers
B	1.542 nanometers
C	0.001491 micrometers
D	1584 picometers

a. Express each of the above measurements in scientific notation with units of meters.

b. Determine the average of the four researchers' results. (Express your answer in scientific notation.)

c. How far apart are the largest and smallest reported result? (This is called the "range" of the reported results.)

EXERCISE 23: Suppose a particular bacterium can divide into two bacteria, once each hour. Thus, in one hour, one bacteria becomes two. After two hours, two bacteria can become four (2×2). In three hours, they become eight ($2 \times 2 \times 2$), and so on. If this process continues for a total of 48 hours, how many bacteria will there be? (Write this number both in scientific notation, and in standard notation.)

EXERCISE 24: Three samples are weighed on an electronic balance. The samples weigh 0.052476 grams, 0.052482 grams, and 0.052490 grams.

a. Express each weight in scientific notation.

b. What is the difference between the largest and smallest samples? (Express your answer in scientific notation.)

c. What is the prefix that denotes the smallest increment of weight that is reported using this balance (the last digit)?

EXERCISE 25: A 10-gram soil sample is analyzed to determine how many bacteria it contains. The sample is mixed with distilled water to form 100 ml of solution. A sample of this solution is diluted to $^1/_{10}$ strength, and 0.1 ml is cultured. From this dilution, a sample is again diluted to $^1/_{10}$ strength (now $^1/_{10} \times {}^1/_{10}$ of full strength) and cultured. This process is continued until a culture is obtained that produces individual colonies (that is, starting with individual bacterium). The culture from the ninth dilution produces 23 colonies.

a. What is the strength of the ninth dilution, as a decimal fraction (expressed in scientific notation) of the strength of the initial soil solution?

b. If 23 bacteria were present in 0.1 ml of the ninth dilution, how many bacteria were present in the 100 ml of the ninth dilution?

c. Divide the number of bacteria in the ninth dilution (determined in Part b) by the strength of the ninth dilution (determined in Part a) to arrive at an estimate of the number of bacteria present in the initial soil sample. (Express this number in scientific notation.)

INDUSTRIAL TECHNOLOGY

EXERCISE 26: Cosmic rays that enter the earth's atmosphere from outer space typically have an energy of about 2 GeV, or 2 giga-electron volts. Express the energy of these rays in electron volts (eV) using scientific notation.

EXERCISE 27: The spark plugs in most automotive engines must provide a spark for every two revolutions of the engine. While driving at normal speeds, the engine may be running at about 2500 revolutions per minute (rpm).

a. If an average speed of 1 mile per minute is assumed for an annual mileage of 10,000 miles, approximately how many minutes is the car driven during the year?

b. At 2400 revolutions per minute, or 1200 sparks per minute, during the year's driving, about how many times does the spark fire during the year? (Express your answer in scientific notation.)

EXERCISE 28: For relatively low temperatures, a thermocouple made with lead and gold wires produces 2.90 microvolts for each degree Celsius (using 0°C as the reference).

a. Express the voltage as volts per degree Celsius in scientific notation.

b. What voltage would you expect from a thermocouple experiencing a temperature of 15°C? (Express your answer in scientific notation.)

EXERCISE 29: A certain isotope of radioactive uranium has a half-life of 4.5×10^9 years. This means that after one half-life, the sample will have one-half of the beginning level of radioactivity.

a. What is the half-life of this isotope in standard notation?

b. If a generation is considered to be 40 years, how many generations will it take for this isotope's radioactivity to be reduced by half?

EXERCISE 30: Radio and television frequencies are given in *hertz*, or cycles per second. Listed below are some frequencies for other common forms of electromagnetism. Use the prefixes to convert each frequency to scientific notation with units of Hz.

Type broadcast	Frequency
Household electricity	60 Hz
AM radio	1080 kHz (kilohertz)
Short-wave radio	10 MHz (megahertz)
FM radio	102 MHz (megahertz)
Radar	8 GHz (gigahertz)
Microwave communication	12 GHz (gigahertz)
Visible light	400 THz (terahertz)

EXERCISE 31: A quartz crystal is commonly used in a digital watch to provide a stable time base. A certain crystal oscillates with a frequency of 5.0 MHz (megahertz). One hertz (Hz) is equivalent to one vibration or cycle per second.

a. Express the quartz frequency of vibration in scientific notation, with units of hertz.

b. With such a frequency, very small time divisions are possible. The period of one cycle can be computed by calculating $^1/_{\text{frequency}}$. How many seconds is the smallest time division using this quartz crystal? (Express your answer in scientific notation.)

EXERCISE 32: A common measure applied to solutions is its pH—a measure of the hydrogen-ion activity of a solution. The hydrogen activity of pure water at 25°C is 0.0000001 moles per liter. A highly acidic solution has 1.0 mole per liter of hydrogen activity, while a highly basic solution has an activity of 0.00000000000001 moles per liter.

a. Express each of the three hydrogen activities given above in scientific notation.

b. The pH value of a solution is simply the exponent of ten of the measure of its hydrogen activity. What is the pH value associated with the highly acidic solution above? with the pure water? with the highly basic solution?

EXERCISE 33: A computer is advertised as having a processing speed of "11 mips," or 11 million instructions per second.

a. Express this speed in scientific notation.

b. On the average, how long does it take to process each instruction at such a speed?

c. How many "nanoseconds" is this?

EXERCISE 34: Steel is very strong, but can be stretched when a stress is applied. From a table of the "modulus of elasticity" for various materials you can find that steel has a modulus of 30×10^6 pounds per square inch. The elongation (or lengthening because of a stress) of a steel beam can then be calculated by multiplying the length of the beam by the stress (in psi) and dividing by the modulus of elasticity.

a. Convert the length of a 12-foot steel beam to inches.

b. Suppose a stress of 5000 psi is applied to the beam. Compute the elongation of the beam. (Express your answer in scientific notation.)

EXERCISE 35: A furnace used to process aluminum consumes approximately 25,000,000 watt-hours of energy per ton (2000 lb) of aluminum processed.

a. Express the energy figure above in scientific notation.

b. If a furnace processes 3200 tons of aluminum during a given period, about how much energy is used? (Express your answer in scientific notation.)

c. Energy is usually reported in kilowatt-hours (kWh) rather than watt-hours. Express the answer to Part b in terms of kWh.

EXERCISE 36: An elevator with a mass of 1200 kg (2640 lb) is lifted upward at a constant speed of 2.0 meters per second.

a. Multiply the mass of the elevator by 9.8 newtons per kg to find the force needed to lift the elevator. (Express your answer in scientific notation.)

b. Multiply the force by the speed to find the power (in watts) needed to lift this elevator at this speed.

c. Power is commonly reported in kilowatts (kW). How many kilowatts of power are needed to lift this elevator?

EXERCISE 37: Electricity travels at very near the speed of light, 3.0×10^8 meters per second. A surge suppressor—used to protect computer equipment for example—advertises a 1-ns clamping time. That is, it can suppress a voltage spike in 1 nanosecond.

a. Express the advertised clamping time in scientific notation, with units of seconds.

b. Multiply the speed of the electricity in a wire by the clamping time to estimate how far along the wire (in meters) a voltage spike will travel before being brought under control by the surge suppressor.

EXERCISE 38: A warehouse module is to be equipped with exhaust fans. The module is 500 feet wide and 700 feet long, and has a ceiling that is 25 feet high. Each fan can move 500 cubic feet per minute. An air exchange of 3 times per hour is desired (that is, the volume of air in the module needs to be replaced 3 times each hour).

a. What is the volume of the warehouse module, in cubic feet? (Express your answer in scientific notation.)

b. What is the volume of air that should be exchanged each hour? (Express your answer in scientific notation.)

c. What volume would this be each minute?

d. How many of the 500-cfm fans are needed to move this air?

e. What effect would material stored inside the module have on this calculation?

EXERCISE 39:

Security coding, for example of automatic garage door openers, is done by setting a bank of switches. Since the number of possible switch settings is so great, it is assumed that the likelihood of an intruder accidentally guessing the setting is very small. For example, with three such switches that can be set to ON or OFF, the number of possible settings is $2 \times 2 \times 2$, or 8 settings. Some switches have three possible settings, indicated as $+1$, 0, or -1, yielding a total of $3 \times 3 \times 3$, or 27 possible settings for three switches.

a. How many possible settings are in a switch bank of 8 switches with three settings for each? (Express your answer in scientific notation.)

b. How many possible settings are in a bank of 12 switches with three settings for each? (Express your answer in scientific notation.)

EXERCISE 40:

A wire-wound resistor with a resistance of 1 ohm (1 Ω) is needed. You have a supply of 8-gauge, 24-gauge, and 36-gauge copper wire that has a resistivity of 1.72×10^{-8} Ω•m. The cross-sectional area of the 8-gauge wire is 8.367×10^{-6} m^2, of the 24-gauge wire is 2.048×10^{-7} m^2, and of the 36-gauge is 1.267×10^{-8} m^2.

a. Compute the resistance per meter of each gauge wire by dividing the resistivity by its cross-sectional area.

b. Use the resistance per meter computed in Part a to determine what length of wire would be needed to obtain the desired resistance of 1 Ω, for each wire gauge.

GLOSSARY

Base

The number that is used as a factor a given number of times. How many times it is used as a factor is indicated by an exponent. For example, in the number 10^3, 10 is the **base**.

Exponent

The number above and to the right of the base number that tells how many times the base is used as a factor. For example, in 10^3, 3 is the **exponent**.

Factors

The numbers multiplied together to form a product. For example, in the multiplication, $4 \times 8 = 32$, the factors are 4 and 8; the product is 32.

Power of ten

The power to which ten is raised in a particular instance. For example, 0.001 is equal to *ten to the negative third power* (or 10^{-3}) and 10,000 is equal to *ten to the fourth power* (or 10^4).

Scientific notation

A number written in scientific notation is a number expressed as the **product** of a number between 1 and 10 and a power of ten. For example, the number 0.00134 written in scientific notation is 1.34×10^{-3}. The number 980,000 written in scientific notation is 9.8×10^5.